Amit Sachan, Preeti Sachan

Microwave Power Transmission as a Future Feasibilty of Solar Power Satellite

GRIN Verlag

Bibliografische Information der Deutschen Nationalbibliothek:

Die Deutsche Bibliothek verzeichnet diese Publikation in der Deutschen National-
bibliografie; detaillierte bibliografische Daten sind im Internet über http://dnb.d-
nb.de/ abrufbar.

Imprint:

Copyright © 2013 GRIN Verlag GmbH
Druck und Bindung: Books on Demand GmbH, Norderstedt Germany
ISBN: 978-3-656-41438-4

This book at GRIN:

http://www.grin.com/en/e-book/213239/microwave-power-transmission-as-a-future-
feasibilty-of-solar-power-satellite

GRIN - Your knowledge has value

Der GRIN Verlag publiziert seit 1998 wissenschaftliche Arbeiten von Studenten, Hochschullehrern und anderen Akademikern als eBook und gedrucktes Buch. Die Verlagswebsite www.grin.com ist die ideale Plattform zur Veröffentlichung von Hausarbeiten, Abschlussarbeiten, wissenschaftlichen Aufsätzen, Dissertationen und Fachbüchern.

Visit us on the internet:

http://www.grin.com/

http://www.facebook.com/grincom

http://www.twitter.com/grin_com

The Future Feasibility of Solar Power Satellite Is Microwave Power Transmission

Amit Sachan[1], Preeti Sachan[2]

Regional College for Education Research & Technology

ABSTRACT

The search for a new, safe and stable renewable energy source led to the idea of building a power station in space which transmits electricity to Earth. The concept of Solar Power Satellites (SPS) was invented by Glaser in 1968. SPS converts solar energy into microwaves and transmit it to a receiving antenna on Earth for conversion to electric power. The key technology needed to enable the future feasibility of SPS is Microwave Power Transmission. SPS would be a massive structure with an area of about 56 sq. and would, weigh about 30,000 to 50,000 metric ton. Estimated cost is about $74 billion and would take about 30 years for its construction. In order to reduce the projected cost of a SPS suggestions are made to employ extra-terrestrialresources for its construction. This reduces the transportation requirements of such a massive structure. However the realization of SPS concept holds great promises for solving energy crisis.

INTRODUCTION

The new millennium has introduced increased pressure for finding new renewable energy sources. The exponential increase in population has led to the global crisis such as global warming, environmental pollution and change and rapid decrease of fossil reservoirs. Also the demand of electric power increases at a much higher pace than other energy demands as the world is industrialized and computerized. Under these circumstances, research has been carried out to look into the possibility of building a power station in space to transmit electricity to Earth by way of radio waves-the Solar Power Satellites. Solar Power Satellites(SPS) converts solar energy in to micro waves and sends that microwaves in to a beam to a receiving antenna on the Earth for conversion to ordinary electricity.SPS is a clean, large-scale, stable electric power source. Solar Power Satellites is known by a variety of other names such as Satellite Power System, Space Power Station, Space Power System, Solar Power Station, Space Solar Power Station etc.[1].One of the key technologies needed to enable the future feasibility of SPS is that of Microwave Wireless Power Transmission.WPT is based on the energy transfer capacity of microwave beam i.e,energy can be transmitted by a well-focusedmicrowave beam. Advances in Phased array antennas and rectennas have provided the building blocks for a realizable WPT system [2].

WHY SPS

Increasing global energy demand is likely to continue for many decades. Renewable energy is a compelling approach – both philosophically and in engineering terms. However, many renewable energy sources are limited in their ability to affordably provide the base load power required for global industrial development and prosperity, because of inherent land and water requirements. The burning of fossil fuels resulted in an abrupt decrease in their .it also led to the green house effect and many other environmental problems. Nuclear power seems to be an answer for global warming, but concerns about terrorist attacks on Earth bound nuclear power plants have intensified environmentalist opposition to nuclear power. Moreover, switching on to the natural fission reactor, the sun, yields energy with no waste products. Earth based solar panels receives only a part of the solar energy. It will be affected by the day & night effect and other factors such as clouds. So it is desirable to place the solar panel in the space itself, where, the solar energy is collected and converted in to electricity which is then converted to a highly directed microwave beam for transmission. This microwave beam, which can be directed to any desired location on Earth surface, can be collected and then converted back to electricity. This concept is more advantageous than conventional methods. Also the microwave energy, chosen for transmission, can pass unimpeded through clouds and precipitations.

SPS-A GENERAL IDEA

Solar Power Satellites would be located in the geosynchronous orbit.the difference between existing satellites and SPS is that an SPS would generate more power-much more power than it requires for its own operation. The solar energy collected by an SPS would be converted into electricity, then into microwaves. The microwaves would be beamed to the Earth's surface, where they would be received and converted back into electricity by a large array of devices known as rectifying antenna or rectenna.(Rectification is the process by which alternating electrical current ,such as that induced by a microwave beam , is converted to direct current). This direct current can then be converted to 50 or 60 Hz alternating current [4]. Each SPS would have been massive; measuring 10.5 km long and 5.3 km wide or with an average area of 56 sq.km.The surface of each satellite would have been covered with 400 million solar cells. The transmitting antenna on the satellite would have been about 1 km in diameter and the receiving antenna on the Earth's surface would have been about 10 km in diameter [5].The SPS would weigh more than 50,000 tons. The reason that the SPS must be so large has to do with the physics of power beaming. The smaller the transmitter array, the larger the angle of divergence of the transmitted beam. A highly divergent beam will spread out over a large area, and may be too weak to activate the rectenna.In order to obtain a sufficiently concentrated beam; a great deal of power must be collected and fed into a large transmitter array.

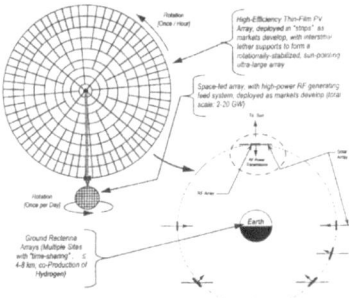

Figure 1 Configuration of SPS is space.

The day-night cycle ,cloud coverage , atmospheric attenuation etc.reduces the amount of solar energy received on Earth's surface.SPS being placed in the space overcomes this .Another important feature of the SPS is its continuous operation i.e,24 hours a day,365 days a year basis. Only for total of 22 in a year would the SPS would be eclipsed for a period of time to amaximum of 72 min.If the SPS and the ground antenna are located at the same longitude, the eclipse period will center around midnight [7]. The power would be beamed to the Earth in the form of microwaves at a frequency of 2.45 GHz. Microwaves can pass unimpeded through clouds and rain .Microwaves have other features such as larger band width , smaller antenna size, sharp radiated beams and they propagate along straight lines. Because of competing factors such as increasing atmospheric attenuation but reducing size for the transmitting antenna and the other components at higher frequency , microwave frequency in the range of 2-3 GHz are considered optimal for the transmission of power from SPS to the ground rectenna site[7].A microwave frequency of 2.45 GHz is considered particularly desirable because of its present uses for ISM band and consequently probable lack of interference with current radar and communication systems. The rectenna arrays would be designed to let light through, so that crops or even solar panels could be placed underneath it. Here microwaves are practically nil [4]. The amount of power available to the consumers from one SPS is 5 GW.the peak intensity of microwave beam would be 23 mW/cm^2.So far, no non thermal health effects of low level microwave exposure have been proved, although the issue remains controversial [4]. SPS has all the advantage of ground solar, plus an additional advantage; it generates power during cloudy weather and at night. In other words SPS receiver operates just like a solar array. Like a solar array, it receives power from space and converts it into electricity. If the satellite position is selected such that the Earth and the Sun are in the same location in the sky, when viewed from the satellite, same dish could be used both as solar power collector and the microwave antenna. This reduces the size and complexity of satellite [8]. However, the main barrier to the development of SPS is social, not technological. The initial development cost for SPS is enormous and the construction time required is very long. Possible risks for such a large project are very large, pay-off is uncertain. Lower cost technology may be developed during the time required to construct the system. So such a large program requires a step by step path with immediate pay-off at each step and the experience gained at each step refine and improve the risk in evolutionary steps [9].

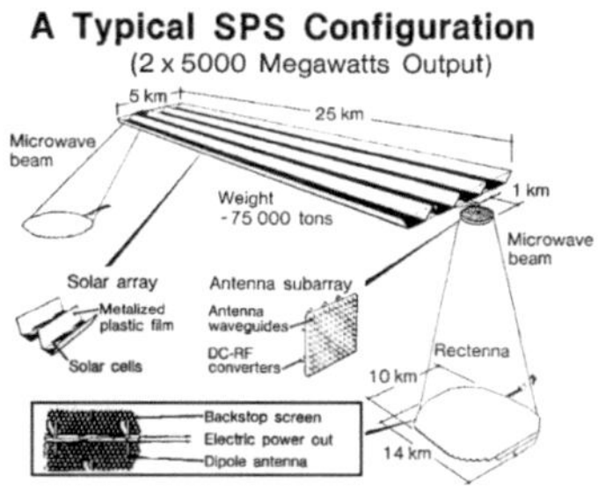

Figure 2

WIRELESS POWER TRANSMISSION

Transmission or distribution of 50 or 60 Hz electrical energy from the generation point to the consumer end without any physical wire has yet to mature as a familiar and viable technology.However, the reported works on terrestrial WPT have not revealed the design method and technical information and also have not addressed the full-scale potential of WPT as compared with the alternatives, such as a physical power distribution line [10]. However the main thrust of WPT has been on the concept of space-to-ground (extraterrestrial) transmission of energy using microwave beam.

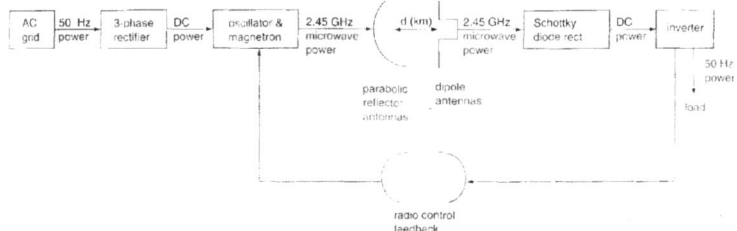

Figure 3 -- conceptual model for a WPT system annexed to a grid.

The 50 Hz ac power tapped from the grid lines is stepped down to a suitable voltage level for rectification into dc. This is supplied to an oscillator fed magnetron. Inside the magnetron electrons are emitted from a central terminal called cathode. A positively charged anode surrounding the cathode attracts the electrons. Instead of traveling in a straight line, the electrons are forced to take a circular path by a high power permanent magnet. As they pass by the resonating cavities of the magnetron, a continuous pulsating magnetic field i.e., electromagnetic radiation in microwave frequency range is generated. After the first round of cavity-to-cavity trip by the electrons is completed the next one starts, and this process continues as long as the magnetron remains energized. Fig.4 shows the formation of a re-entrant electron beam in a typical six cavity magnetron. The output of the rectifier decides the magnetron anode dc voltage. This in turn controls the radiation power output. The frequency of the radiation is adjusted by varying the inductance or capacitance of the resonating cavities.

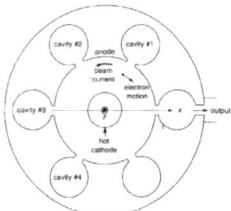

Figure 4. Re-entrant electron beam in a six-cavity magnetron

The microwave power output of the magnetron is channelled into an array of parabolic reflector antennas for transmission to the receiving end antennas. To compensate for the large loss in free space propagation and boost at the receiving end the signal strength as well as the conversion efficiency, the antennas are connected in arrays. Moreover, arrayed installation of antennas will necessitate a compact size. A series parallel assembly of schottky diodes, having a low standing power rating but good RF characteristics is used at the receiving end to rectify the received microwave power back into dc. Inverter is used to invert the dc power into ac. A simple radio control feedback system operating in FM band provides an appropriate control signal to the magnetron for adjusting its output level with fluctuation in the consumers demand at the receiving side. The feedback system would switch of the supply to the oscillator and magnetron at the sending end if there is a total loss of load.

The overall efficiency of the WPT system can be improved by

> ➢ Increasing directivity of the antenna array
> ➢ Using dc to ac inverters with higher conversion efficiency
> ➢ Using schottky diode with higher rating.

>
MICROWAVE POWER TRANSMISSION IN SPS

The microwave transmission system as envisioned would have had three aspects [5]:

1. The conversion of direct power from the photovoltaic cells, to microwave power on the satellites on geosynchronous orbit above the Earth.
2. The formation and control of microwave beam aimed precisely at fixed locations on the Earths surface.
3. The collection of the microwave energy and its conversion into electrical energy at the earth's surface.

The ability to accomplish the task of efficiently delivering electrical power wirelessly is dependent upon the component efficiencies used in transmitting and receiving apertures and the ability to focus the electromagnetic beam onto the receiving rectenna. Microwave WPT is achieved by an unmodulated, continuous wave signal with a band width of 1Hz. Frequency of choice for microwave WPT has been 2.45GHz due to factors such as low cost power components, location in the ISM band, extremely low attenuation through the atmosphere [2]. The next suggested band centered at 5.8GHz system reduces the transmitting and receiving apertures. But this is not preferred due to increased attenuation on higher frequency. The key microwave components in a WPT system are the transmitter, beam control and the receiving antenna called rectenna .At the transmitting antenna, microwave power tubes such as magnetrons and klystrons are used asRF power sources. However, at frequencies below 10 GHz, high power solid state devices can also be used. For beam safety and control retro directive arrays are used. Rectenna is a component unique to WPT systems. The following section describes each of these components in detail.

TRANSMITTER

The key requirement of a transmitter is its ability to convert dc power to RF power efficiently and radiate the power to a controlled manner with low loss. The transmitter's efficiency drives the end-to-end efficiency as well as thermal management system i.e., any heat generated from inefficiencies in the dc-RF conversion, should be removed from the transmitter as it reduces the life time of RF devices and control electronics [2]. Passive inter modulation is another field which requires critical attention. Filtering of noise and suppression of harmonics will be required to meet he regulatory requirement. The main components of a transmitter include dc-to-RF converter and transmitting antenna. . The complexity of the transmitter depends on the WPT application. For the large scale WPT application such as SPS, phased array antennas are required to distribute the RF power sources across the aperture and electronically control the power beam. Power distribution at the transmitting antenna=$\sqrt{(1-r^2)}$, where r is the radius of antenna [7]. There are mainly three dc-to-RF power converters: magnetrons,klystrons and solid state amplifiers.

Klystron

Fig. shows the schematic diagram of a klystron amplifier [15].

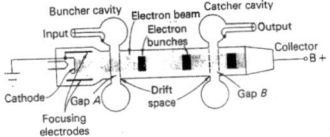

Here a high velocity electron beam is formed, focused and send down a glass tube to a collector electrode which is at high positive potential with respect to the cathode. As the electron beam having constant velocity approaches gap A, they are velocity modulated by the RF voltage existing across this gap. Thus as the beam progress further down the drift tube, bunching of electrons takes place. Eventually the current pass the catcher gap in quite pronounce bunches and therefore varies cyclically with time. This variation in current enables the klystron to have significant gain. Thus the catcher cavity is excited into oscillations at its resonant frequency and a large output is obtained.

Fig.6 shows a klystron transmitter [2]. The tube body and solenoid operate at 300°C and the collector operates at 500°C. The overall efficiency is 83%. The microwave power density at the transmitting array will be 1 kW/m² for a typical 1 GW SPS with a transmitting antenna aperture of 1 km diameter. If we use 2.45 GHz for MPT, the number of antenna elements per square meter is on the order of 100. Therefore the power allotted to the individual antenna element is of the order of 10 W/element. So we must distribute the high power to individual antenna through a power divider [1].

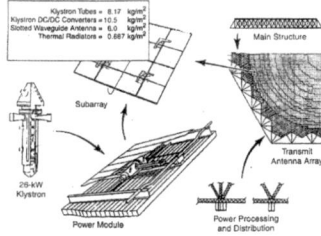

Figure 6 Klystron transmitter

BEAM CONTROL

A key system and safety aspect of WPT in its ability to control the power beam. Retro directive beam control systems have been the preferred method of achieving accurate beam pointing. As shown in fig.7 a coded pilot signal is emitted from the rectenna towards the SPS transmitter to provide a phase reference for forming and pointing the power beams [2]. To form the power beam and point it back forwards the rectenna, the phase of the pilot signal is captured by the receiver located at each sub array is compared to an onboard reference frequency distributed equally throughout the array. If a phase difference exists between the two signals, the received signal is phase conjugated and fed back to earth dc-RF converted. In the absence of the pilot signal, the transmitter will automatically dephase its power beam, and the peak power density decreases by the ratio of the number of transmitter elements.

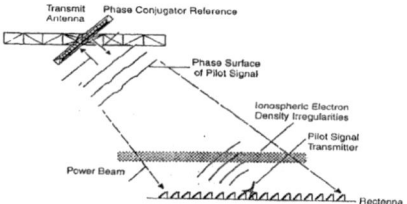

Figure 7 Retro directive beam control concept with an SPS.

RECTENNA

Brown was the pioneer in developing the first 2.45GHz rectenna [2]. Rectenna is the microwave to dc converting device and is mainly composed of a receiving antenna and a rectifying circuit. Fig .8 shows the schematic of rectenna circuit [2]. It consists of a receiving antenna, an input low pass filter, a rectifying circuit and an output smoothing filter. The input filter is needed to suppress re radiation of high harmonics that are generated by the non linear characteristics of rectifying circuit. Because it is a highly non linear circuit, harmonic power levels must be suppressed. One method of suppressing harmonics is by placing a frequency selective surface in front of the rectenna circuit that passes the operating frequency and attenuates the harmonics.

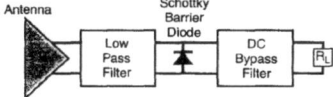

Figure 8 Schematic of rectenna circuit.

For rectifying Schottky barrier diodes utilizing silicon and gallium arsenide are employed. In rectenna arrays, the diode is the most critical component to achieve higher efficiencies because it is the main source of loss. Diode selection is dependent on the input power levels. The breakdown voltage limits the power handling capacity and is directly related to series resistance and junction capacitance through the intrinsic properties of diode junction and material .For efficient rectification the diode cut off frequency should be approximately ten times the operating frequency. Diode cut off frequency is given by $f = 1/[2\pi R_s C_j]$, where f is the cut off frequency, R_s is the diode series resistance, C_j is the zero-bias junction capacitance.

RECENTLY DEVELOPED MPT SYSTEMS

The Kyoto University developed a system called Space Power Radio Transmission System (SPORTS) [1]. The SPORTS is composed of solar panels, a microwave transmitter subsystem, a near field scanner, a microwave receiver. The solar panels provide 8.4 dc power to the microwave transmitter subsystem composed of an active phased array. It is developed to simulate the whole power conversion process for the SPS, including solar cells, transmitting antennas and rectenna system. Another MPT system recently developed by a team of Kyoto University ,NASDA and industrial companies of Japan , is an integrated unit called the Solar Power Radio Integrated Transmitter (SPRITZ),developed in 2000 [1]. This unit is composed of a solar cell panel, microwave generators, transmitting array antennas and a receiving array in one package. This integrated unit as shown in fig.9 could be a prototype of a large scale experimental module in the orbit.

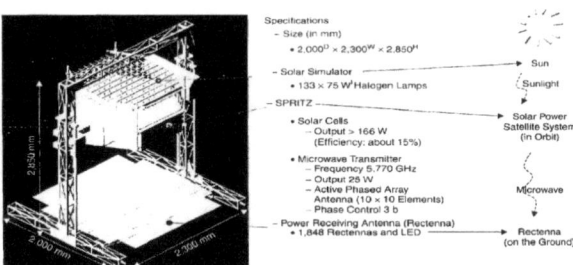

Figure 9 SPRITZ (Solar Power Radio Integrated Transmitter 2000)

CONSTRUCTION OF SPS FROM NON TERRESTRIAL

MATERIALS: FEASIBILITY AND ECONOMICS

SPS, as mentioned before is massive and because of their size they should have been constructed in space [5]. Recent work also indicate that this unconventional but scientifically well –based approach should permit the production of power satellite without the need for any rocket vehicle more advanced than the existing ones. The plan envisioned sending small segments of the satellites into space using the space shuttle. The projected cost of a SPS could be considerably reduced if extra-terrestrial resources are employed in theconstruction [9].One often discussed road to lunar resource utilization is to start with mining and refining of lunar oxygen, the most abundant element in the Moon's crust, for use as a component of rocket fuel to support lunar base as well as exploration mission. The aluminum and silicon can be refined to produce solar arrays [12]. A number of factors combine to make the concept of using non conventional materials appear to be feasible. Among them are the shallow gravity wells of the Moon and asteroids; the presence of an abundance of glass, metals and oxygen in the Apollo lunar samples; the low cost transport of those materials to a higher earth orbit by means of a solar-powered electric motor; the availability of continuous solar energy for transport, processing and living [12]. Transportation requirement for SPS will be much more needed for known for known commercial applications. One major new development for transportation is required: the mass driver [12].The mass driver is a long and narrow machine which converts electrical energy into kinetic energy by accelerating 0.001 to 10 kg slugs to higher velocities. Each payload-carrying bucket contains superconducting coils and is supported without physical contact by means of dynamic magnetic levitation. As in the case of a linear synchronous motor-generator, buckets are accelerated by a magnetic field, release their payload, decelerate with return energy and pick up another pay load for acceleration. The power source can be either solar or nuclear. The mass driver conversion efficiency from electrical to kinetic energy is close to 100 percent. The mass driver can be used as a launcher of lunar material into free space or as a reaction engine in space, where payloads are transferred from orbit to orbit in a spiral trajectory. The performance of the mass driver could match that of the space shuttle main engines. But the mass driver has the advantage that any material can be used as fuel and continuous solar power in space is the common power source.An alternative to the use of lunar resources for space manufacturing is the use of earth-approaching asteroidal materials.

MICROWAVES-ENVIRONMENTAL ISSUES

The price of implementing a SPS includes the acceptance of microwave beams as the link of that energy between space and earth. Because of their large size, SPS would appear as a very bright star in the relatively dark night sky. SPS in GEO would show more light than Venus at its brightest. Thus, the SPS would be quite visible and might be objectionable. SPS posses many environmental questions such as microwave exposure, optical pollution that could hinder astronomers , the health and safety of space workers in a heavy-radiation (ionizing) environment , the potential disturbance of the ionosphere etc.The atmospheric studies indicate that these problems are not significant , at least for the chosen microwave frequency [13]. On the earth, each rectenna for a full-power SPS would be about 10 km in diameter. This significant area possesses classical environmental issues. These could be overcome by siting rectenna in environmentally insensitive locations, such as in the desert, over water etc. The classic rectenna design would be transparent in sunlight, permitting growth and maintenance of vegetation under the rectenna. However, the issues related to microwaves continue to be the most pressing environmental issues. On comparing with the use of radar, microwave ovens , police radars, cellular phones and wireless base stations, laser pointers etc. public exposures from SPS would be similar or even less. Based on well developed antenna theory, the environmental levels of microwave power beam drop down to $0.1\mu W/cm^2$ [12]. Even though human exposures to the 25 mW/cm^2will, in general, be avoided, studies shows that people can tolerate such exposures for a period of at least 45 min. So concern about human exposure can be dismissed forthrightly [4]. Specific research over the years has been directed towards effects on birds, in particular. Modern reviews of this research show that only some birds may experience some thermal stress at high ambient temperatures. Of course, at low ambient temperatures the warming might be welcomed by

8

birds and may present a nuisance attraction [13]. Serious discussions and education are required before most of mankind accepts this technology with global dimensions. Microwaves, however is not a 'pollutant' but , more aptly , a man made extension of the naturally generated electromagnetic spectrum that provides heat and light for our sustence.

ADVANTAGES AND DISADVANTAGES

The idea collecting solar energy in space and returning it to earth using microwave beam has many attractions.
1. The full solar irradiation would be available at all times expect when thesun is eclipsed by the earth [14]. Thus about five times energy could be collected, compared with the best terrestrial sites
2. The power could be directed to any point on the earth's surface.
3. The zero gravity and high vacuum condition in space would allow much lighter, low maintenance structures and collectors [14].
4. The power density would be uninterrupted by darkness, clouds, or precipitation, which are the problems encountered with earth based solar arrays.
5. The realization of the SPS concept holds great promises for solving energy crisis
6. No moving parts.
7. No fuel required.
8. No waste product.

The concept of generating electricity from solar energy in the space itself has its inherent disadvantages also. Some of the major disadvantages are:

1. The main draw back of solar energy transfer from orbit is the storage of electricity during off peak demand hours [15].
2. The frequency of beamed radiation is planned to be at 2.45 GHz and this frequency is used by communication satellites also.
3. The entire structure is massive.
4. High cost and require much time for construction.
5. Radiation hazards associated with the system.
6. Risks involved with malfunction.
7. High power microwave source and high gain antenna can be used to deliver an intense burst of energy to a target and thus used as a weapon[15].

CONCLUSION

The SPS will be a central attraction of space and energy technology in coming decades. However, large scale retro directive power transmission has not yet been proven and needs further development. Another important area of technological development will be the reduction of the size and weight of individual elements in the space section of SPS. Large-scale transportation and robotics for the construction of large-scale structures in space include the other major fields of technologies requiring further developments. Technical hurdles will be removed in the coming one or two decades. Finally, we look forward to universal acceptance of the premise the electromagnetic energy is a tool to improve the quality of life for mankind. It is not a pollutant but more aptly, a man made extension of the naturally generated electromagnetic spectrum that provides heat and light for our sustenance. From this view point, the SPS is merely a down frequency converter from the visible spectrum to microwaves.

REFERENCES

[1] Hiroshi Matsumoto, "Research on solar power satellites and microwave power transmission in Japan", IEEE microwave magazine, pp.36-45, Dec 2002.

[2] James O. Mcspadden & John C. Mankins,"Space solar power programs and microwave wireless power transmission technology", IEEE microwave magazine, pp.46-57, Dec 2002.

[3] J.C. Mankins,"A fresh look at space solar power: new architectures, concepts and technologies" in 38th Astronautical Federation.

[4] Seth Potter, "Solar power satellites: an idea whose time has come [online] Available on www.freemars.org/history/sps.html, last updated on Dec.1998

[5] Consumer Energy Information: EREC Reference Briefs [online] Available on www.eere.gov/consumerinfo/rebriefs/123.html,last updated on Apr.03.

[6] Mc GrawHill Encyclopedia of Science and Technology, vol.16, pp.41.

[7] Om P.Gandhi,"Microwave engineering and application", PHI.

[8] Geoffrey A.Landis,"A super synchronous solar power ", Presented at SPS- 97: Space &electric power for humanity, 24-25 Aug 1997, Montreal, Canada.

[9] Geoffrey A.Landis,"An evolutionary path to SPS", Space power, vol.9, no.4, pp.365-371, 1990.

[10] S.S.Ahmed, T.W.Yeong and H.B.Ahmad,"Wireless power transmission and its annexure to the grid system", IEE Proc.-Gener.Transm.Distrib., Vol.150, No.2, March 2003.

[11]Kennedy "Electronics Communication Systems", Tata McGraw Hill.

[12] B.O'Leary,"The construction of satellite solar power stations from non terrestrial materials: feasibity and economics", Alternative energy sources, Vol.3, pp.1155-1164.

[13] John M.Osepchuk,"How safe are microwaves and solar power from space", IEEE microwave magazine, pp.58-64, Dec.2002.

[14] International Encyclopedia of Energy, Vol.4, pp.771.

[15] David M.Pozar,"Microwave Engineering", Wiley

Mr. Amit Sachan has completed M.Tech degree in department of Electrical Engineering. He continuing his profession as Assistant Professor in Regional College for Education Research & Technology and doing research work in the field of Electrical power Engineering.